CREATION EX NIHILO

It Came From Nothing

///

> "There is an idea--strange, haunting, evocative-one of the most exquisite conjectures in science or religion..an infinite hierarchy of universes, so an elementary particle, such as an electron, would, if penetrated, reveal itself to be an entire closed universe. Within it, organized into the local equivalent of galaxies and smaller structures, are an immense number of other, much tinier elementary particles, which are themselves universes at the next level, and so on forever–an infinite downward regression, universes within universes, endlessly. And upward as well. Our familiar universe of galaxies and stars, planets, and people, would be a single elementary particle in the next universe up, the first step of another infinite regress."
>
> Carl Sagan
>
> ********

I   Is our expanding universe a muon transforming to an electron?

///

Forward

To chronicle creation, is to embrace the notion of space as impedance and gravitation as reactance, for that is the way the world is made. Vacuum divergence is to gravity what acceleration is to inertial matter. Behind it all, a subliminal dynamic. To comprehend the cosmos, one must think inside the box. The box is empty, yet it holds many secrets.

Contents

Forward 5

Introduction 7

A Recipe for Gravity From Inertia 8

Revisiting Inertia as the cause of Gravity 19

Transforming Hubble Energy to infinite Plane 24

The Flat Earth Model 28

///

Introduction

In *Standard Theory*, the influence of separated bodies upon one another relies upon unseen "go-between particles" invented, by their advocates, as momentum transferring entities. The herein developed theory of the void as an inertial dynamic, by contrast, is based upon a hitherto overlooked apotheosis of spatial expansion. In this ambit, all things are spatially connected.

If it costs nothing to create a mass as proclaimed by Richard Feynman, then creation must include a "**g**" field befitting each mass.[1] As to be shown, '**g**' fields themselves are instantiated phenomenon(s) emerging as they do, from the confluence of three cosmological concurrences----> dynamic expansion of inertial space ----> 2nd law reactance "**ma**" -----> cosmic counter action. Within what is known of the extent of space and the mass of matter, the cosmos is a superlative vacuum, far beyond that attainable in laboratories. Yet the inertia of the universe is present in some fashion to oppose accelerating motion.[2] By what manner can space bring about gravity, inertial reaction and the anomalies of propagation?

Surprisingly, the existence of such an artifice has long been known, but commonly viewed as having little cosmological applicability – indeed, the miraculous properties of infinite laminae(s) were considered largely academic. That all changed in the latter years of the 20th century. While Edwin Hubble's work had long previously demolished Einstein's static universe, it wasn't until the 1998 supernova studies, that the imperatives of exponential expansion and flat space of infinite extent, would be faced. Coming with this new knowledge, were the apparent problem of finding sufficient energy to fund the acceleration process.

The initial object will be that of determining the effective area density of the infinite plane that corresponds to the density of a sample volume taken as the Hubble sphere. It's a two step process: 1) transformation of Hubble mass to surface density and, 2) finding the density factor that corresponds to the infinite plane equivalent. While neither Hubble mass nor size is separately utile as an extrapolation base for an infinite universe, taken together as area density, they supply the missing physical factors upon which the Laws of inertia and gravity were derived *in absentia*

As an operative geometry, the infinite plane fulfills the job description. That the ratio of Hubble mass to Hubble area is approximately *one/kg per square meter* suggests the universe is a *"put up job."*[3] Yet when examined in the light of modern methods, we see a universe governed by necessity. Apparent alternatives are illusory, there is no evidence of providential fine tuning nor is there reason to invoke the anthropic principle. The focus here is to introduce inertial space and derive its properties. Gravitational force (local '**g**' fields and the value of **G**) follow straightaway thereafter. As suspicioned by Richard Feynman, gravity turns out to be an incorrectly perceived pseudo force. But to complete the specification of '**g**' fields as inertial reactions, Feynman's musings would require an isotropic acceleration field. Long in coming, homage is hear paid to the 1998 empiricist(s) - whose work will now be acclaimed as foundational to understanding gravity in terms of expansion. Proposed more than a century past, the cosmological constant Λ can now be seen as the great irony of 20th century physics. That the critical value $\Lambda = 3H^2$ (adapted by Einstein to prevent gravitational collapse), in the end, turns out to be the root cause of gravity.

---

[1] Feynman, Lectures on Gravity

[2] The principle relating other mass to local inertia is attributed to the 19th century physicist, Ernst Mach.

[3] As once quipped by Fred Hoyle in response to the realization that certain cosmological factors appear to be finely tuned to permit the development of life..

A Recipe For Gravity From Inertia

Gravity requires three ingredients (cosmological expansion, spatial inertia and a measured unit M_B of reactionary mass-energy). For expansion, the exponential parameter c^2/R is mandated as natures prescription for zero energy self instantiated cosmology. Blend then, spatial inertia, not as volumetric density ρ_U but as area density σ_U from the ratio of Hubble to mass to area $M_U/(4\pi R^2)$:

$$G = \frac{c^2}{4\pi R \sigma_U}$$

The value of the spatial inertia factor σ_U appears at first glace, a put up job. As an organic property of the cosmos relating different forms of energy within a volume defined by a closed surface, its root has been a mystery since first formalized by Isaac Newton as the 2nd law of motion. Expanding inertial space creates isotropic momentum divergence. The reactance of matter to isotropic momentum divergence is called gravity. To express gravity in terms of spatial expansion, it is necessary to first determine the inertial property of space σ_U.

A perfect vacuum, by definition, is empty of mass. In its 3-D form as thinly distributed lumps of high density matter, space is commonly reckoned as a perfect fluid having nil resistance to motion. Average density $\rho_U = 3 \times 10^{-26}$ kg/m³ (an unattainable vacuum by laboratory standards). If however, the universe is constructed from a rigid material (depicted as a cube **U** in **Fig 1**), it would exhibit inertial mass in the range of 1.5×10^{53} kg. How is it, a volume substantially devoid of matter, enlists the inertial content of the cosmos, to inaugurate a counter force - dependent only upon the mass (ostensibly independent of the volume, area, shape or uniformity of the accelerated object)?

Fig 1 depicts a Hubble sized cubical volume **U** sampled from infinite flat space (having a side length $L \approx 10^{26}$ meters and density $\rho_U \approx 3 \times 10^{-26}$ kg/m³). In contact with face f_1 of **U**, a solid rigid cube **B** having cross sectional area **A** and mass M_B. If U is accelerated (red arrow) or **B** is accelerated (blue arrow), the force felt on f_1 will depend upon the structure of **U**. If constituted from a rigid material, the pressure at the interface f_1 would be $Mu(a)/A$. But as thinly distributed clumps of matter, **U** offers nil inertial resistance, limited to masses actually in contact with **B**. Mysteriously, the universe strikes a middle course. From an inertial perspective *a la* Newton's 2nd law, the cosmos functions neither as a vacuum nor a solid, but as an area density. To assess this peculiarity, we consider mass M_U spread uniformly over the area of **U**. Any body metaphorically entering he universe would be opposed by the area density of a face. Within the universe an accelerating body is opposed as if the masses of the two faces were added and then shared by any number of infinite parallel planes distributed there over. Thus for accelerations between an infinite universe and a finite **B**, the reactive force is supplied by the combined area density of two parallel faces $f_1 + f_3 = M_U/3L^2$. With the knowledge that fields created by area densities will be $\perp$ to the surface, the force between '**B**' and **U** can be expressed in terms of a common contact area, e.g. one meter². Then:

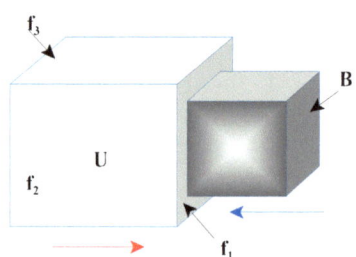

If **B** is accelerated to the left (blue arrow) at rate a_1, the force upon f_1 is $M_B[a_b]/3$.

If **U** is accelerated to the right (red arrow) at rate a_u, the force upon **B** is $M_U[a_u]/3$

Within a [1 m²] common area of contact, there is an acceleration '**g**' that satisfies the equation:

$$[g][M_U/meter^2] = [a_1][M_B/meter^2] \qquad (1)$$

From (1) the inertial mass of **U** has meaning in relation to **B** only when M_U is amalgamated with space to create an area density. Inasmuch as there are three directions to space, the inertia of the universe can be modeled as three perpendicular planes (double density faces as shown in **Fig 2**). Each of the three mutually orthogonal planes then contains 1/3 the total energy M_U (and as explained below) represents a composite density comprised of a plurality of parallel flat slabs normal to each spatial direction.[4] By this intrigue, the volumetric density of the cube is divided into 3 separate directional densities each organized as a plurality of parallel flat slabs. A cube of side "**L**" has volume L^3 and three orthogonal surfaces areas L^2, so for the three planes:

$$M_U = \rho_U(L^3) \quad \text{and} \quad M_U = \sigma_U(3L^2) \quad (2)$$

Whence
$$\sigma_U = \rho_U(L^3)/(3L^2) = \rho_U(L/3) \quad (3)$$

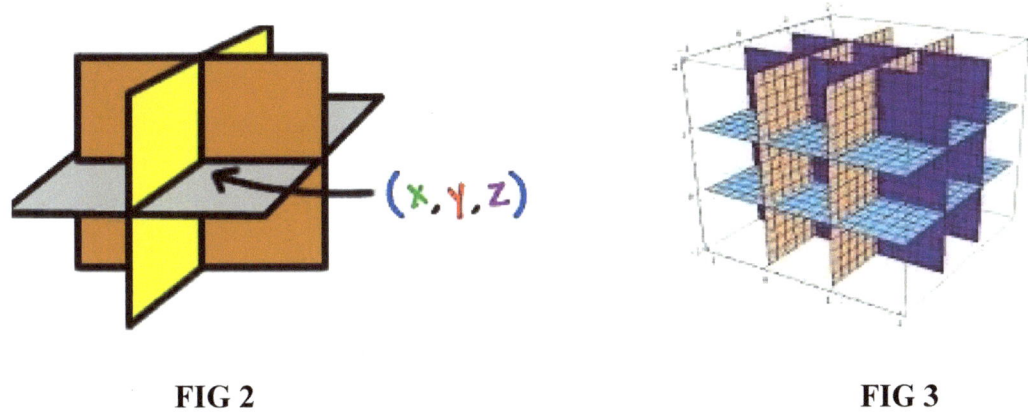

FIG 2   **FIG 3**

That the universe can be functionally described in terms of three orthogonal planes (**Fig 2**), it can also be modeled as a plurality of parallel planes as shown if **Fig 3**. In truth, the operative essence of space as a dynamic reactance, reduces to a realization that the inertial functionality of space as a dynamic impedance does not derive from the ostensible form of space as a vacuous volume lightly sprinkled with matter. Cosmic opposition to acceleration is not the manifest of a three dimensional aggregate, but rather as a plurality of parallel area densities. To extend **Fig 3** to a panoptic emulation, the volume of the cosmic sample must be strati graphed as a plenum of infinite area planes. Reckoned from the perspective of an accelerating mass, however, reactance emerges in form as a single infinite area plane orthogonal to the instantaneous direction of acceleration.

Since cosmic mass is engaged in opposing acceleration, curiosity arises as to what is gained by reckoning opposition to acceleration as a set of orthogonal planes as opposed to single volume? That they are operatively equivalent to a single plane, adds to the wonder of the resolution.

---

[4]From a Theorem due to Gauss, the integral of a vector divergence field over the volume containing the divergence equals the integral over the surface that contains the volume. By this formalism the density per side is $\rho_U(L/6)$ but density per direction will be $\rho_U(L/6)$. Mass M_U is considered uniformly distributed throughout the volume and the same mass is considered uniformly distributed over three orthogonal surfaces.

Cosmic inertial response to acceleration is instantaneous, as if the universe were transformed into a ubiquitous area density plane of infinite extent. The inertial response of the thinly sliced similitude of planes function collectively rather than locally. The miraculous property of infinite planes, the reason why the universe acts as it does, follows from the fact that field lines are ($\perp$) to the surface. Ergo, the influence of an ∞ plane does not diminish with distance. Illustrated in **Fig 4**, (whether it be an electrically charged surface or a gravity field arising from a mass density per unit area), will be independent of distance. Thus g_1 will equal g_2 and the intensity of gravitational field will be same at any height. Consequently, all parallel planes can be considered a single plane -- if they have the same area density they will exert the same influence upon distant matter as if they were in contact therewith. The inertial influence of all planes can be merged into a single plane perpendicular in contact area with an accelerating body '**B**' per **Fig 1**. From the perspective of '**B**,' the universe appears as a confronting area density, instantly manifesting as garnered from its volumetric density.

The above posed query re the operative functionality of the cosmos as laminae, resolves with relentless cogency. Spatial inertia is the mean between inertia imposed by rigidity of a structured solid and the rarity of the real cosmos as a near empty vacuum. By natures mathematical ruse, the perpendicularity of the field forces created by ∞ planes augment additively. By this fortune we are alive *a la* the goldilocks cosmic state.

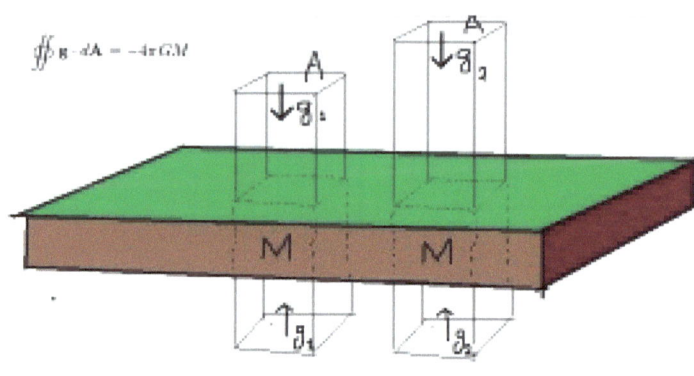

FIG 4

Newton discovered the Law, later researchers adjusted the force unit to comport with '*one-kgm-meter-per-second-squared.*' In doing so, they were defining the inertial modulus of the universe in terms of force, mass and time. Area did not enter into the prescription. Why is it that neither the area of an accelerated mass nor the area of the universe, appear in Newton's 2nd Law?

Because the universe acts as area density independent of location, an accelerating body having irregular shape and non uniform density, will nonetheless feel cosmological impedance as the sum of infinitesimally thin stratifications, each having its own density map which projects upon the area density of the cosmic plane. The geometric(s) are illustrated in **Fig 5 and Fig 6**.

When the flat square sheet in **Fig 5** is accelerated normal to its surface, the projection on the cosmic area density plane is a square, at an inclined angle, the flat sheet experiences the inertia of the universe as a rectangle. When accelerated parallel to an edge, the cosmos is seen as a line. The body shown in **Fig 6**, will, for accelerations in different directions, cast different shapes upon cosmic inertial plane. The length of the black lines indicate the strength of the cosmological reactionary force for the stratified projections in each direction of acceleration $\perp$ to the sides of the object

Fig 7, a body **B** of mass **M** is typified by atoms (red) joined by springs. Acceleration in the horizontal direction will compress horizontal springs whereas vertical acceleration will compress vertical springs. Total inertial reaction, however, will be the same in either case, because all atoms are involved in any acceleration irrespective of direction.

FIG 5

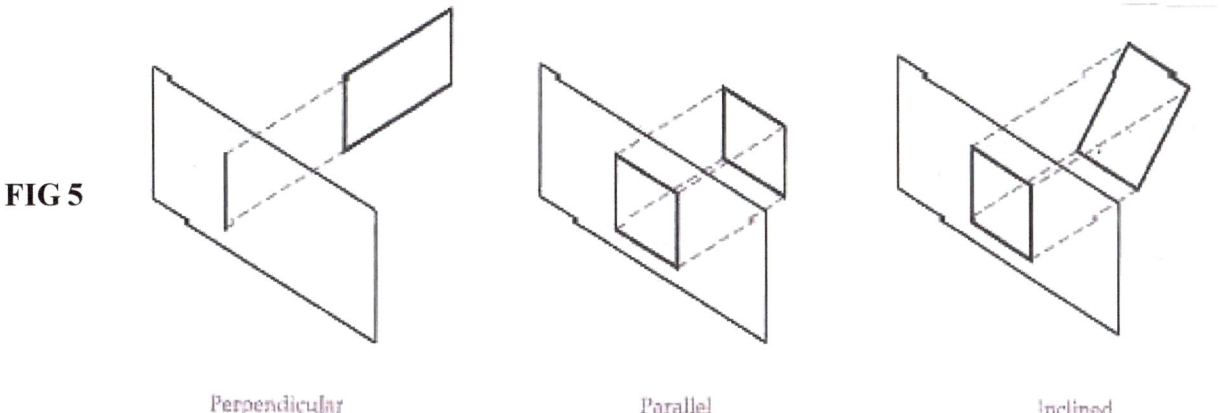

Perpendicular   Parallel   Inclined

FIG 6

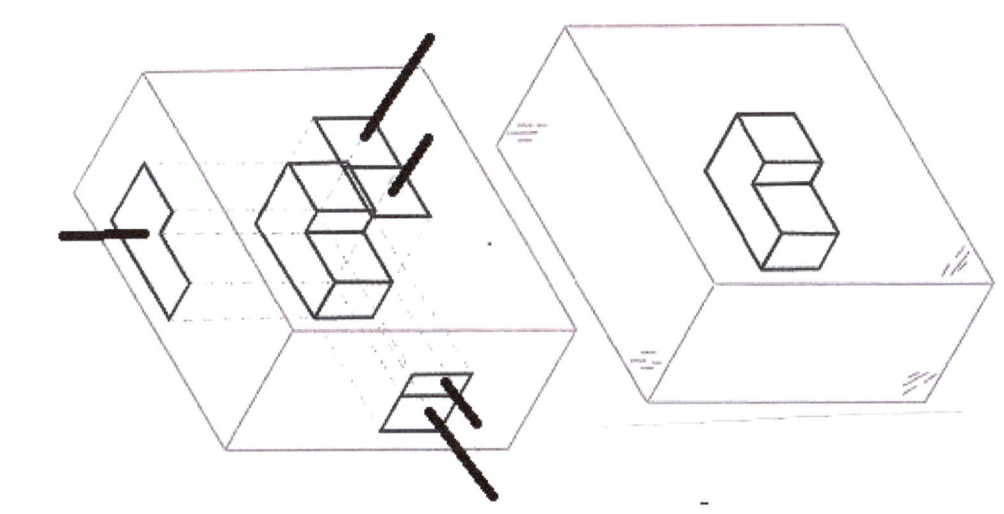

FIG 7

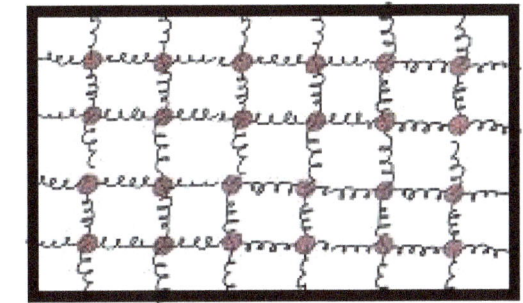

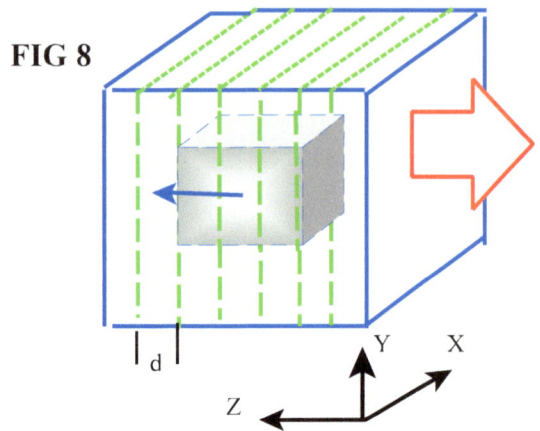

FIG 8

Fig 8 shows the block **B** inside **U**. As previous indicated in (2) and (3), a homogeneous volume can be simulated as a laminae of area densities (dotted green). Since the inertial effect of an infinite plane is independent of distance, the inertial affect of the cube appears operationally equivalent to a single plane having a composite area density equal to the total mass of the cube divided by area of the effective surface orthogonal to the direction of acceleration. Inertial pressure resulting from acceleration depends only upon the number of mutually exclusive directions in 3-D space. For the cubic sample of the universe **U** there are 6 sides but only 3 degrees of freedom. From (1), for a common area of contact of one square meter:

$$\mathbf{a}_2 \sigma_U = \sigma_B \mathbf{a}_1 \qquad (4)$$

Whence, for unidirectional acceleration of **B** at rate $\mathbf{a}_1$ (blue arrow of **Fig 1** or **Fig 8**) or a unidirectional acceleration of **U** at a rate $\mathbf{a}_2$ (red arrow of **Fig 1** or **Fig 8**) the effective area density σ_U created by transforming M_U to the operative surface area ($3L^2$) of **U** is $\rho_U(L/3)$. Each of the seven slabs has the same density $\rho_U(L/21)$ and consequently each slab will exert the same inertial force throughout the volume of **B**. From (4), it makes no difference whether **U** or **B** is accelerated:

$$\mathbf{a}_2 = \frac{\sigma_B}{\sigma_U} \mathbf{a}_1 \qquad (5)$$

Assuming greater difficulty would arise in attempting to test (5) by accelerating **U**, we focus instead upon the acceleration of **B** and derive $\mathbf{a}_2$ as the reactionary 'g' field of the cosmos due to the acceleration of **B**. In will be understood, the reaction of the universe ($\mathbf{a}_2 = \mathbf{g}$) can be expressed either as an acceleration (m/sec^2) or as a force per unit mass (ntn/kg). However viewed, it is defined as the aggregate of the densities presented by that set of planes perpendicular to the direction of acceleration of body **B** relative to the cosmic reference frame. Per (5), reactionary acceleration is inversely proportional to cosmic area density $\rho_U(L/3)$, which for the estimated volumetric density $\rho_U = 3 \times 10^{-26}$ kg/m^3, reduces to one kg/m^2 for a 10^{26} meter cube edge.

In reality, of course, no experiment can be with a single body. Somewhere an equal and opposite force exists - acting upon another mass as required for momentum conservation on the global scale as shown in **Fig 9**. While depicted as separate surfaces it will be understood, the two σ_U planes are metaphorical. There is in reality, one volume (imagined for expletive purposes as divided into a plurality of area density laminae which span the sample volume taken of the universe. Together they create the dynamic inertial factor = one kg/m^2 for both M_{B1} and M_{B2}. In opposing oppositely accelerating masses M_{B1} and M_{B2} the action of the universe is a nothing for nothing momentum exchange - as it must be for a zero energy universe. The cosmos as a whole, is a momentum conserving constellation -- one large fat plane opposing momentum change in any and direction when viewed from the perspective of an accelerating object. By the same token, each object contributes a small part to the σ_U density.

FIG 9:

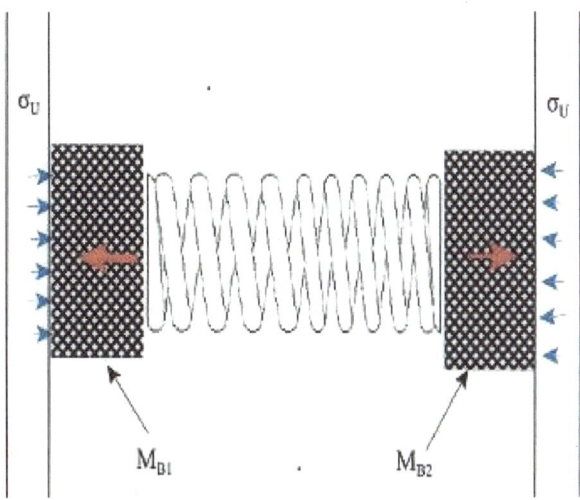

For a body in uniform motion, the universe acts as a superlative vacuum (ρ_U). But when velocity changes in magnitude or direction, the character of the universe abruptly mediates as a counter force. In the guise of inertial reaction, the cosmos can be modeled as a plurality of infinite planes or as one infinite plane density σ_U. Photons and other EM propagation(s) are likewise moderated by σ_U as these forms motion are defined by transverse wave accelerations. In this role, σ_U is expressed as an electrical impedance in terms of inductance μ_o and capacitance ε_o

$$Z = \sqrt{\frac{\mu_o}{\varepsilon_o}} \quad (6)$$

In classical physics, Newton's 2nd law and his law of Gravity both depend upon mass and acceleration. Understanding the one is a condition precedent to understanding of the other.

$$F_I = Ma, \quad g = M\frac{G}{r^2} \quad (7)$$

The first expression states an inertial reactive force F_I arises when a mass M_B is accelerated with respect to the mass $\underline{M}_U^*$ of the universe. The second expression formulates an expression for the inertial reactive **g** field of M_B resulting from the perpetual isotropic acceleration of $\underline{M}_U^*$ relative to M_B. That neither $\underline{M}_U^*$, nor the area, density or shape of M_B appears in the formulation of F_I, would seem to indicate something amiss.[5] The present tendency is to rationalize Newton's 2nd law as a definition that follows from the "*way-the-world-is-made.*" This seeming requires the universe sense the acceleration of each atom and apply the counter force at the instantaneous location of thereof. Because the inertial reactionary force of an accelerated object is ostensibly independent of spatial factors (shape, size, uniformity and orientation relative to the direction of acceleration), they have been routinely dismissed as having no direct bearing upon the outcome.

Beginning with the first expression, we divide both sides of the expression by meters^2, then

$$\frac{F_I}{m^2} = \frac{M_B}{m^2} a_1 \quad (8)$$

Per Newton's 3rd law, inertial force F_I can be expressed in terms of inertial space $\underline{M}_U^*$ x acceleration:

$$a_2 \frac{M_u}{m^2} = \frac{M_B}{m^2} a_1 \quad (9)$$

---

[5] $\underline{M}_U^*$ symbolizes the inertia of empty space. It is the ratio obtained by dividing Hubble mass by Hubble area, herein also denoted as σ_U

Solving for the unknown cosmological acceleration field $\mathbf{a_2}$, then:

$$\mathbf{a_2} = \frac{M_B/m^2}{M_U/m^2}[\mathbf{a_1}] = \frac{\sigma_B}{\sigma_U}[\mathbf{a_1}] \qquad (10)$$

where σ_U and σ_B are area densities expressing the ratios approximated using Hubble parameters as per (5). Taking Newton's well feted 2nd Law as a calculation tool, we determine the value of σ_U based upon an acceleration rate $\mathbf{a_1}$ = one meter/sec^2 for a one kg mass. Then for a one kg sheet having area one square meter accelerated normal to its surface at one meter per sec^2, will experience one ntn reactionary force uniformly distributed over its area. Likewise, the σ_U density plane of the universe will experience one ntn of force over an effective area of one square meter. The acceleration of the mass sheet is communicated to the universe because both the sheet and the acceleration thereof, are embedded therein by virtue of σ_U as the inertial representative thereof.

To complete the development of σ_U as the inertial representation of the universe $\mathbf{U}$, we revert once again to Newtons 2nd law and examine the acceleration of a free body $\mathbf{B}$ of mass M_B with respect thereto. Because the inertial action of the infinite plane can only involve those lines of action normal to the surface which also conjoin M_B, the interaction between $\mathbf{U}$ and $\mathbf{B}$ will be confined to the common area, i.e., the projection of $\mathbf{B}$ upon $\mathbf{U}$. Since σ_U is the infinite plane emulation of cosmic mass in form as area density, the imaginary acceleration opposing force lines will be perpendicular to σ_U. In communicating σ_U as area density acceleration impedance, it is convenient to also express M_B as area density σ_B.[6]

Fig 10 illustrates the mutual domain of action between $\mathbf{U}$ and $\mathbf{B}$ as the cookie cutter area punched out by the projection of $\mathbf{B}$ upon σ_U. In all cases, the void appears to an accelerating object $\mathbf{B}$ as an omnipresent $\perp$ segment of the inertial impedance plane σ_U. In this simile, $\mathbf{B}$ is accelerating downward and the invisible lines emanating from the projection of B upon σ_U are upward.

FIG 10

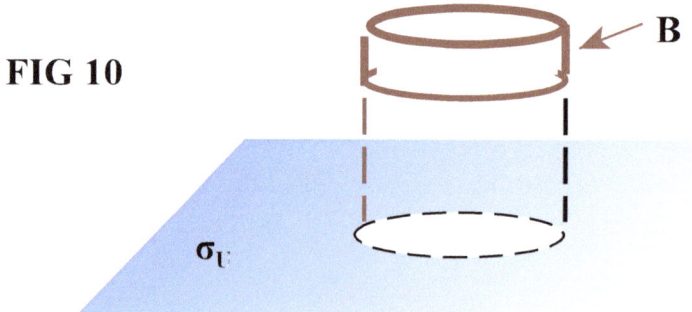

---

[6]That any irregular shape, size or non-uniform density can be transformed into an idealized uniform flat plane area density for purposes of calculating the interaction of the universe thereon, will be recognized is one more adjunct of that property of infinite planes which makes action at a distance independent of the distance.

Applying Newton's 2nd law to the segment defined by the projection of **B** upon σ_U, the normalized expression for reactionary force per unit area in terms of the mass area density of **B** is:

$$\frac{\mathbf{F}}{\mathbf{m}^2} = \frac{\mathbf{M_B}}{\mathbf{m}^2}\mathbf{a} \qquad (11)$$

Force **F**, however, can be expressed as cosmic mass/area multiplied by $\mathbf{a_2}$: with respect to **B**:

$$\frac{\mathbf{M_U}}{\mathbf{m}^2}\mathbf{a_2} = \frac{\mathbf{M_B}}{\mathbf{m}^2}\mathbf{a_1} \qquad (12)$$

Spatial acceleration carries with it the area density characteristic σ_U. When mass is accelerated with respect to space, it encounters the same σ_U density characteristic. An accelerating mass feels the pressure $[(\sigma_U)\mathbf{a_1}]$ as momentum flow. Force corresponding to the rate of change in the velocity of $\mathbf{M_B}$ is countered by momentum flow influx (pressure) arising from acceleration of $\mathbf{M_B}$ wrt to σ_U. Writing (12) in terms of σ_U and σ_B, then,

$$\frac{\mathbf{F}}{\mathbf{m}^2} = \mathbf{a_2}\sigma_U = \mathbf{a_1}\sigma_B \qquad (13)$$

For an acceleration $\mathbf{a_1}$ of one meter per second squared, and with mass $\mathbf{M_B}$ spread over a uniform flat disk of area one square meter, as shown in **Fig 11**, we can determine σ_U based upon the fact that acceleration of the universe $\mathbf{a_2}$ wrt to **B** equals the acceleration of **B** wrt to σ_U. If the cross section area of **B** is one square meter, then since **F** equals one ntn when a one kg mass is accelerated at one meter per sec^2, σ_U pressure equals one ntn/m^2. Consequently:

$$P = \sigma_U \times (\mathbf{1\ m/sec^2}) \qquad (14)$$

Witness whereof, (14) can only be true if σ_U = one kg/m^2. That it will have the same numerical value irrespective of the mass, shape, size or acceleration of **B**, the cosmic scalar density plane σ_U can be regarded as the "*connective constant*" between the inertial property of the universe as a whole and the inertial mass of individual elements that constitute the whole. Newton's 2nd law follows, beautiful in its simplicity, obscuring as it does, one of natures great sublimities σ_U.[7] As a modifier it has no influence upon the force, appearing as it does as a unity factor in all expressions thereof.[8]

---

[7]The 17th Century experimenters following Newton's discovery of his laws of motion, combined the dimensionality(s) of space, time and mass to define the Newtonian unit of force [1 ntn = 1 kg per meter per second squared]. The cause of inertia as distant matter, although never formulated into a testable theory, had been suggested by several historical figures. As an interface between the whole and its parts, appears to have escaped interrogation.

[8]Provisionally, σ_U may be considered temporally invariant. During expansion, cosmic volume expands equally in all directions. The laminae opposing acceleration in a particular direction, increase in area as the square of the Hubble radius **R** whereas the number of laminae increase proportionate with Hubble scale **R**. The effective particle density thus diminishes inversely. But as to be shown, particle inertia increases in proportion to **R**, hence area inertial density remains constant.

To compute reactionary pressure from cosmological parameters, multiply the *area normalized mass* [M_B/A_B] of an accelerating mass M_B by its acceleration a_1 wrt the rest frame of the universe and divide by σ_U. Both M_U and M_B are then expressed in terms of a common area density of one square meter. By this means, the physical reality of the universe as a "plenum of planes" of infinite extent is operatively convened within the ambit of an interface region having an area defined by M_B as though it were spread uniformly to create an area density of one kg/m².

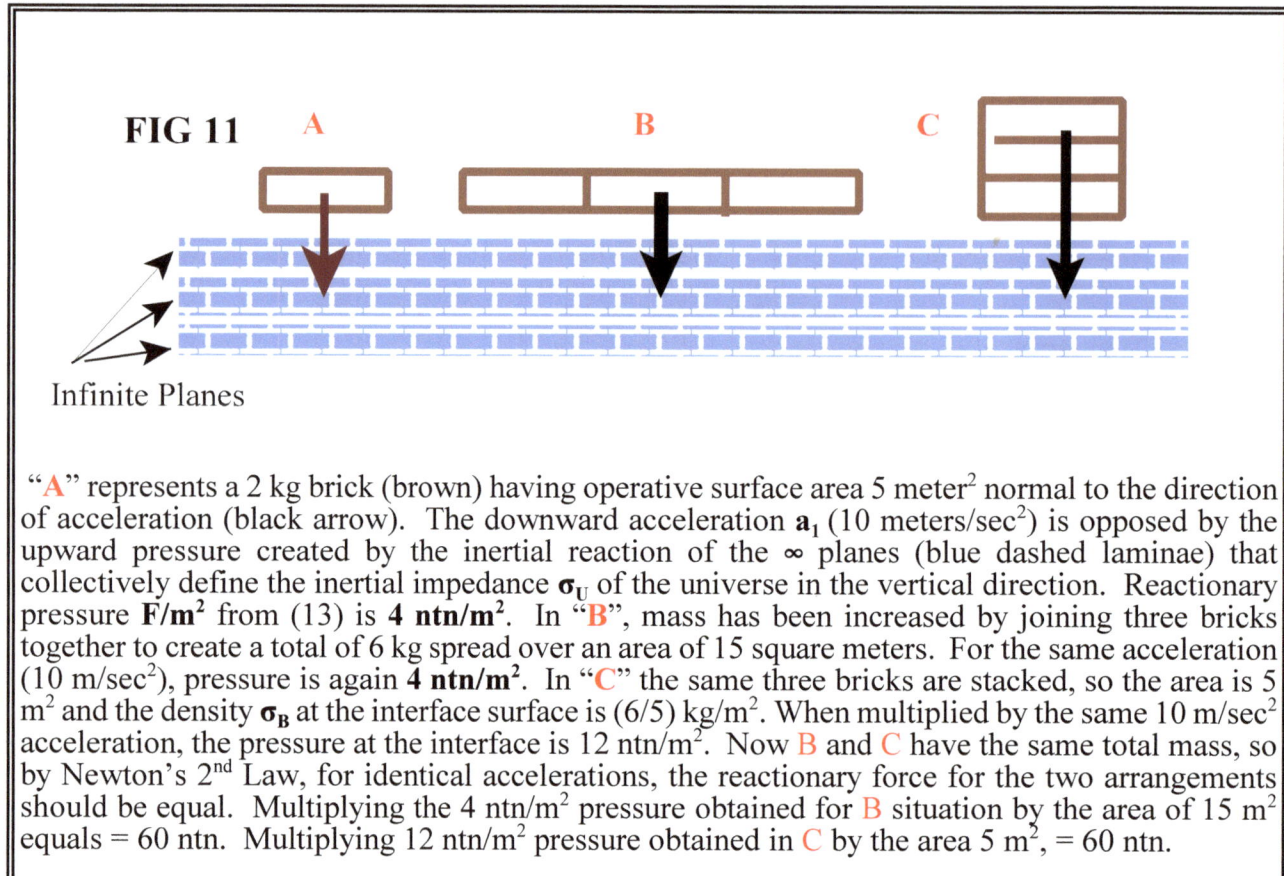

FIG 11

Infinite Planes

"A" represents a 2 kg brick (brown) having operative surface area 5 meter² normal to the direction of acceleration (black arrow). The downward acceleration a_1 (10 meters/sec²) is opposed by the upward pressure created by the inertial reaction of the ∞ planes (blue dashed laminae) that collectively define the inertial impedance σ_U of the universe in the vertical direction. Reactionary pressure F/m^2 from (13) is **4 ntn/m²**. In "B", mass has been increased by joining three bricks together to create a total of 6 kg spread over an area of 15 square meters. For the same acceleration (10 m/sec²), pressure is again **4 ntn/m²**. In "C" the same three bricks are stacked, so the area is 5 m² and the density σ_B at the interface surface is (6/5) kg/m². When multiplied by the same 10 m/sec² acceleration, the pressure at the interface is 12 ntn/m². Now B and C have the same total mass, so by Newton's 2nd Law, for identical accelerations, the reactionary force for the two arrangements should be equal. Multiplying the 4 ntn/m² pressure obtained for B situation by the area of 15 m² equals = 60 ntn. Multiplying 12 ntn/m² pressure obtained in C by the area 5 m², = 60 ntn.

σ

Fig 11, and the accompanying description, illustrate the principle of inertial continuity. Irrespective of size, shape or oddity, all accelerated bodies receive equal treatment from the universe. That pressure is momentum flow, it will be conserved in all transformations between the universe and its parts. Cosmic inertial density σ_U, in rendering as a plurality of planes, suggests the efficacy of the laminae construct as a model for its contents. By this ruse, the inertial-gravitational forces acting upon any object, are reduced to a simple area density ratio between the object and the universe. Consistent with the development of inertial reaction, the formulations (4), (5) and (13), we now apply the same rationale to calculate gravitational force. From Gauss's law, the expression for the acceleration of the universe in \terms of its area density is

$$4\pi\sigma_U G = c^2/R = a_n \tag{15}$$

For **B** is expressed as an area density

$$4\pi\sigma BG = a_2 = g. \tag{16}$$

From (15),

$$G = \frac{c^2}{4\pi R \sigma_U} \quad (17)$$

Combining (15) and (16)

$$\sigma_U g = \sigma_B(a_n) \quad (18)$$

To test the inertial theory of gravity upon a spherical body of known mass and size, we imagine the earth tortured into a flat disk having an area commensurate with its spherical surface per **Fig 12**. Taking the earths mass as 5.98×10^{24} kg and radius as 6.37×10^6 meters, the area density of the disk is approximately:

$$\sigma_E = \frac{M_E}{4\pi (r_e)^2} = \frac{5.98 \times 10^{24} \text{ kg}}{4\pi (6.37 \times 10^6 \text{ meters})^2} = 1.173 \text{ kg/m}^2 \quad (19)$$

From (18), the reactionary pressure created by spatial expansion $[a_n = c^2/R]$ at earths surface is:

$$g(\sigma_U) = (\sigma_E)a_n = (1.173 \text{ kg/m}^2)c^2/R = (1.173 \text{ kg/m}^2)[(3 \times 10^8 \text{ m/sec})^2/(1.1 \times 10^{26} \text{ m}) = 9.8 \text{ nn/m}^2$$

Whence: $g = (9.8 \text{ ntn/m})/\sigma_U = 9.8 \text{ m/sec}^2 \quad (20)$

FIG 12

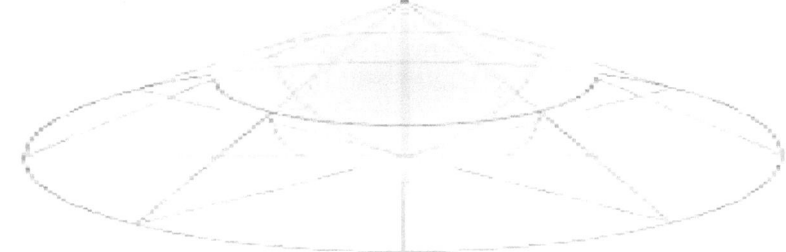

In summary, *Gravity* can be regarded as the inertial reaction of matter created by volumetric spatial expansion. Non expanding matter (objects having mass) obstruct the uniformity of the expansion field., creating stress tension between space and mass. Stress takes form as negative pressure, i.e., momentum flow from σ_U to the individual masses that occupy the volume. In the above example, momentum flow from the universe enters across the earths surface (meaning the two dimensional boundary which includes the geosphere, hydrosphere and what is covered thereby).

While the stress created between expanding space and non-expanding matter is commonly referred to as the earths '**g**' field, the strength of the field is, in actuality, the collective sum of the pseudo force fields of each interior atom or inertial entity. The action of the cosmological acceleration factor upon the surface density σ_B will be equal to the action of the reactionary acceleration of the earth multiplied by counter field of the universe.

Inertia has long been suspected to be somehow determined by global mass. A primary hurdle for this proposition, is that the affect of cosmic mass should not modify the inertial coefficient '**M**' by which inertia is reckoned as a participant in the expression for momentum (**Mv**), kinetic energy (**M**v^2/2) and a myriad other formulations in which **M** plays a definitive role. An additional obstacle to the idea that local inertia depends upon distant matter, is that, by all accounts, inertial reaction is instantaneous. This was anathema to Einstein. The dependence of instantaneous inertial reaction upon far away influences was reasoned to require "*instantaneous-action-at-a-distance.*"

But as functional opposition to acceleration in the form of an infinite plane, the distance between the plane and the accelerating body, is of no consequence. Reactionary fields are not maintained nor updated by mythical *speed-of-light* particles. The inertial modulus σ_U is ubiquitous and omnipresent. Gravitational and electrical force fields associated with masses and charges are spatial extensions of the particle - constituting as they do, the essence of the entity. Gravity due to inertia has its analogy in electric field due to charge. Neither is autonomous -- they have no separate existence therefrom.

The imposition of the heretofore unrecognized unity factor σ_U as arrant to an understanding of 2nd law dynamics, will indeed come as surprise to those practiced in modern methods of instruction. To understand Newton's 2nd law in terms of Mach's Principle is to comprehend '**g**' fields as inertial reaction to global expansion.

Newton's 2nd Law and the consequence thereof (local '**g**' fields) are the pearls of Newtonian physics. Gravitational fields are (as Richard Feynman suggested more that a half century past), improperly perceived pseudo forces - brought about by inertial reaction of masses stressed by global expansion. The acceleration factor required to convert Newton's 2nd Law of Motion into his Law of Gravity corresponds to the cosmological expansion rate. The underlying spatial dimensionality needed to merge Newton's two Laws is obscured by the simplicity of its reduced form **F = Ma**. It is, however, a common denominator to the way the universe works - as pressure a la momentum flow, and as density per the inertial factor intrinsic to spatial accelerations in both the expression of force as pressure and the expression of mass as density. Being common to both sides of the equation the common area factor cancels leaving no vestigial evidence (a caveat to those who pursue reduction as the ultimate goal). The spatial commonality factor, while redundant from a mathematical perspective, is key to understanding Newtonian inertia. In the reduced expression known as the 2nd Law of motion, it has vanished. But in this treatise, the concern is with origins. The common practice of reducing equations to their simplest form, can sometimes erase valuable source information. In the case of inertia (and its reactionary '**g**' field powered by global expansion), the resolution was obscured and consequently long overlooked.

In Einstein's words, a theory should be as simple as possible, but not simpler. The introduction of spatiality clarifies the physiology. The means by which infinite planes operate to resist peculiar forms and shapes is demystified. With the revelation of the *modus operandi*, the universe becomes a bit more intelligible, and an old principle is revived. Mach's Principle will be recognized as fully embraced within the definition of σ_U[9] [Conveying as it does the inertia of the universe diluted by the area thereof, it brings together the ubiquitous imposition of spatial impedance to any change in momentum]. Gravity follows as σ_U reactance to isotropic spatial divergence.

---

[9] Although Mach was not the first to propose the dependence of local inertial upon cosmic mass, It is now accredited to him because Einstein, during his development of General relativity, frequently referred to the concept as Mach's Principle.

Part II

Revisiting Inertia as Gravitational Cause

Fig 13 (A, B, C, D, E): The adaption of flat plane geometrics restated and illustrated.

Fig 13-A: Construct begins with a uniform density cube. Then $F \approx M_B M_A G/d^2$ per Newton's Law of gravity. If it is desired to increase the force by adding additional blocks, where is the most advantageous location to place them? In **Fig 13-B**, identical cubes are laid in a row parallel to the **X** axis. Because each cube is further away by distance '**d**,' the affect upon **B** of successive cubes is reduced by the square of the distance(s), $(2d)^2, (3d)^2, (4d)^2, (5d)^2$. Adding more cubes directly along the **X** line of action per **Fig 13-B**, increases force, but because '**g**' fields fall off inverse square with distance, the added force diminishes quickly (no matter how many cubes are added, total force F_T is less than twice that produced by A_1 alone, that is $[F_T < [2G \times 10^{-26} M_B/d^2]$.[10] Suppose instead, cubes are added along the **Y** axis as shown in **Fig 13-C**. This creates a $\perp$ column. Each new cube adds a small force at a less favorable angle, and because it is also further away from **M**, the effort appears to be even less productive than adding cubes along the line of action -- that is, until the number of cubes becomes large. When this occurs, the lines of force associated with the orthogonal column are perpendicular to the column. The '**g**' field from gauss's Law

$$\int_S \mathbf{g} \cdot \mathbf{n} \, dA = -4\pi Gm \tag{21}$$

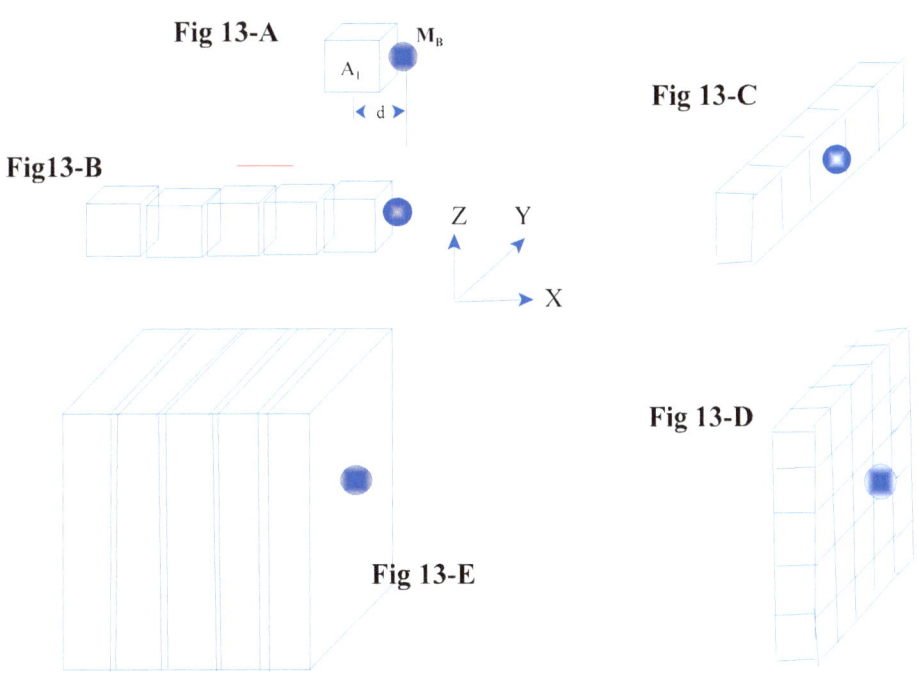

---

[10]Known as the Basel Problem, originally proposed by Mengoli in 1644 and solved by Euler in 1734:

$$\sum_{n=1}^{n=\inf inity} \frac{1}{n^2} = \frac{1}{1^2} + \frac{2}{2^2} + \ldots \ldots \frac{1}{n^2} = \frac{\pi^2}{6}$$

For an infinite orthogonal column, the density of the force is determined using an imaginary cylindrical Gaussian enclosure of radius 'd' and length L placed to encompasses the cubical column concentric therewith. Assuming the column has a linear mass density, the flux is normal to the surface of the cubes, consequently, none escapes through the ends of the cylinder. Integration over the area of the cylinder gives:

$$-g(2\pi d)L = -4\pi Gm \qquad (22)$$

Since **M** is the total mass enclosed by the surface **S**, then for a segment of length **L**, and density λ:

$$g = 2G\lambda/d \qquad (23)$$

The '**g**' field falls off inversely with distance rather than inverse squared (per **Fig 13-B**).

Suppose while building orthogonally in the **Y** direction, cubes are also added in the **Z** direction to construct a slab (**Fig 13-D**) parallel to the **YZ** plane, and consequently $\perp$ to the **X** axis. The slab has a mass to area density σ, the object then being to calculate the '**g**' field at distance '**d**' from the center of the slab. In this case, the appropriate Gaussian surface is a pillbox (**Fig 10**), whose flat end areas "A_P" are normal to the plane. Acceleration is (parallel to the outward normal unit vector), so the sides contribute nothing to the flux integral. The area integral will be twice the end area, hence:

$$-g\int_S dA = -4\pi Gm = -g(2A_p) \qquad (24)$$

The mass enclosed by **S** is a cookie cutter punch of area A_p and density σ, so it has mass σA_p. Therefore:

$$-g(2A_p) = -4\pi G\sigma A_p \qquad (25)$$

Whence:
$$g = 2\pi G\sigma \qquad (26)$$

For the infinite slab, gravitational acceleration is independent of the distance from the slab. By constructing a second, third and forth infinite slab parallel to the first as in **Fig 13-E**, the force upon M_B is increased proportionally. Since each infinite slab exerts the same force upon M_B irrespective of the perpendicular distance from the wall, all slabs parallel to the **YZ** plane augment equally to increase the '**g**' force upon M_B. By this contrivance, multiple **YZ** slabs can be functionally considered as a single infinite plane of density $2\pi Gn\sigma$ in virtual contact with any mass contained therein. As previously elaborated, inertial reaction is instantaneous since every slab makes its presence upon M_B continuously known. The density function of the universe is the same in any direction - the effect of all area density slabs presents a ubiquitous orthogonal scalar opposition to acceleration irrespective of heading. For an accelerating object, the cosmos appears as inertial plane. In reality, of course, the division of the cosmic block into slabs is metaphorical -- there are no individual slabs nor is there a single slab. The conceptual implementation of an infinite plane as a dimensional reducing simile can be appreciated in the context of what causes spatial inertia. As previously developed, Newton's 2nd law is now explainable in terms of his 3rd law, and the physiology by which cosmic content is marshaled to oppose acceleration, is made intelligible.

$$F = M_U a_1 = M_B a_2 = \frac{M_U}{\text{meter}^2}g = \frac{M_B}{\text{meter}^2}a_2 \qquad (27)$$

On the large scale, the universe is homogeneous and isotropic. Thus for a sphere such as the Hubble:

$$\int_V \rho_U (dV) = \int_S \sigma_U (ds) \tag{28}$$

The integral over the volume is $(4/3)\pi R^3$ and the integral over the surface is $4\pi R^2$. Hence:

$$\sigma_U = \rho_U R/3 \tag{29}$$

The Hubble sphere, in its function as an observational limit, offers nil resistance to acceleration in its form as a three dimensional density. Herein we have developed the modus operandi of the universe as an infinite plane to explain how nearly empty space impedes changing momentum. Velocities are relative, but peculiarly mensurable by the difference in the rate at which time is accumulated by relatively moving clocks. Changing velocity is relative to the universe. Historically, the mystery of inertia has been reduced to one of two alternatives: Either:

1) Mass-energy is intrinsically endowed with an acceleration impeding quality, or,
2) Matter and space coalesce as a unified global inertial property of the cosmos.

That the inertial property of universe can be defined in terms of Hubble parameters as a functional ensemble of infinite planes, is a compelling endorsement of latter alterative. The physiology of any single plane can be imagined as the positive bare mass energy contained in a Hubble sized sample of the universe, spread uniformly over a flat plane having the same surface area as the Hubble sphere. Said surface density -- by the principle of cosmic homogeneity, will be the same everywhere.

Infinite planes as extensively reiterated herein, have unique properties. They exert the same gravitational and inertial influence everywhere. Consequently, one infinite plane is mathematically sufficient to specify the operative inertia of the universe. An infinite plane exerts the same gravitational force upon a mass irrespective of its distance from the plane. While the density is determined by a sample size, any size will be satisfactory, so long as it is sufficiently large to obtain a good estimate of average density. The area density of a plane formed by spreading the estimated mass of the Hubble sphere over a surface commensurate with its manifold, is in the range of one kilogram per square meter. That it is exactly one kg per square meter is a result of the definitions established between force, mass and acceleration.

Infinite plane mechanics emerges as the inertial operative of large volumes –> the inevitable consequence of the conservation laws of momentum and energy. There is no law of "conservation of mass." In our universe, momentum x mass, is the fundamental entity that is conserved. Any change in momentum must be countered instantly by the reaction of inertial space

<u>Conservation of momentum requires large homogenous volumes take effect as area densities.</u>
In the infinite plane model of flat inertial space, cosmic mass is amalgamated within a spatial surface area to create an area density σ_U. The plane so formed is but a similitude, yet it produces real physical results in form as instantaneous spatial impedance to momentum change anywhere.

A second known property of space, is that it expands at an accelerating rate a_n. These two factors (σ_U and a_n), when multiplied together, define the vacuum pressure $-P$. When expressed as a ratio, they define G. Specifically:

$$-P = (\sigma_U)(c^2/R) \quad \text{And} \quad a_n/\sigma_U = 4\pi G \tag{30}$$

A mass or charge spread over a large flat area is sometimes called a Bouguer plate. The field created thereby is perpendicular to the surface area, consequently the intensity of the field is constant over the surface area and independent of the distance therefrom. For a uniform distribution of mass, however, there can be no gravity field and no force without acceleration. That this requirement is satisfied by exponential expansion of inertial space, we are alive to perceive the universe as the emolument thereof. In keeping with the long standing superstition of local '**g**' forces as an emanation of gravitons, it should be sobering to reflect upon the fact that an acceleration factor **G** must be introduced into both the Newtonian and Einsteinian formulations of gravity in order to vivify inert matter as a force. Specifically, per (17) and (30), **G** incorporates both the inertial area density σ_U and the exponential expansion rate c^2/R.

The area density plane σ_U defines inertial force in terms of relative acceleration. Masses act in two capacities -- immersed as they are as individual inertial entities scattered throughout inertial space, any momentum thereof, is resisted by the collective inertia σ_U of the whole. However, as inertia entities, each mass is also a subscribing contributor to σ_U. As such, all are embedded within σ_U as part of inertial space. In opposing the acceleration of other bodies, and as a collaborator in the formulation of its own reactionary '**g**' field, each mass creates an inertial reaction, then as a part of the global isotropic expansion field of inertial space each indirectly creates its own counter reactionary '**g**' field. Parallel slabs of mass, produce no gravity field in the space separating the planes, ergo, the affect of a plurality of parallel slabs upon each other is nil. The same is true even though relative acceleration exists between inertial space and individual bodies. Since a plurality of planes can be considered as single omnipresent area density, there is no relative acceleration as between interior planes. Reckoned as an area density operative in the guise of an infinite plane, the universe cannot otherwise be analysized as a volume creating gravity deficit (gravity acting upon gravity). Transformation from ρ_U to a flat σ_U reforms the universe as bare mass inertial area density.

Fig 14

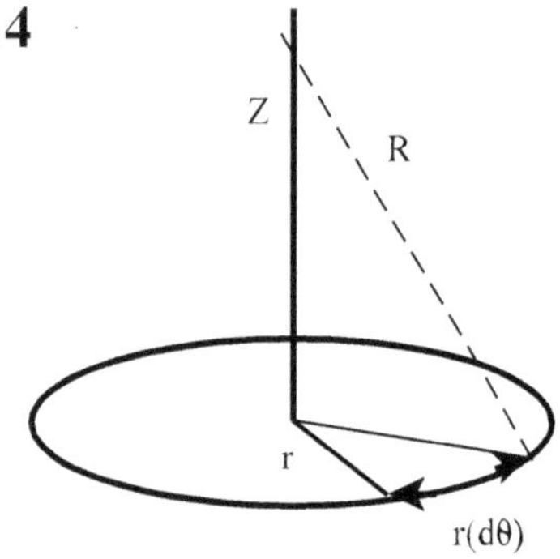

Fig 14 illustrates the geometry for finding the infinite flat plate potential at a point. Consider a disk and let the radius become infinite, then the gravitation potential for a disk of radius '**a**'

$$\Phi = -G\int_S \frac{\sigma(\mathbf{ds})}{R} = 2\sigma\pi G\int_0^a \frac{r(dr)}{(z^2+r^2)^{1/2}}$$

When $r \gg z$, potential will not depend upon '**r**'

$$\Phi = 2\sigma\pi Gz \qquad (31)$$

Fig 15

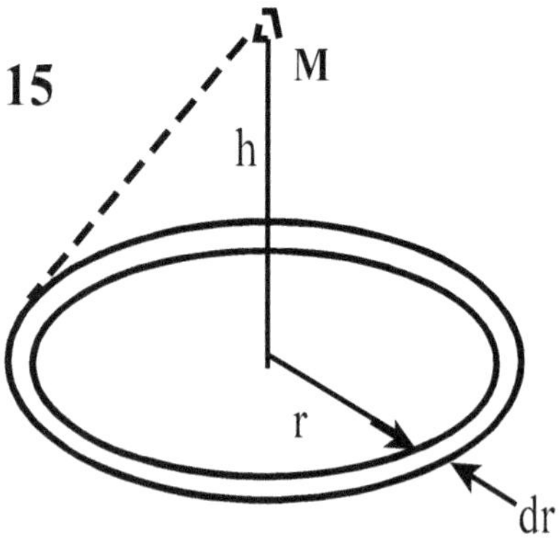

Force lines will be perpendicular to the disk and the imaginary Gaussian surface S is pillbox (a short cylinder whose flat faces of area "A" are parallel to the plane per **Fig 10**. t any height 'h' above the plane, we use **Fig 15** and calculate the gravity acting upon a mass **M** by integrating the force felt by concentric rings from a radius zero to infinity.

$$F/M = 2\pi G\sigma h\int_0^\infty \frac{r}{(h^2+r^2)^{3/2}}\,dr = 2\pi G\sigma$$

Inertia and gravitation are accounted for within the auspicious of a single field, σ_U. Initially constituted from the ratio of Hubble mass to Hubble area, now seen apropos of flat space. In the context thereof, σ_U represents the reactance of the universe. That cosmic mass can be normalized in terms of spatial extent, parlays other analytical rewards, including the catalyzing of **G** from Λ, the explication of Newton's second Law in terms thereof, and the accurate prediction of expansion engendered '**g**' fields emerged therefrom. In the infinite slab model of the universe, the effect of cosmic mass everywhere is made manifest as the inertial opposition to acceleration anywhere. As previously emphasized, Mach's principle is revitalized in the guise of the unity operative σ_U.

With the recognition that flat space reactionary forces are explainable in terms of a large number of parallel slabs acting in concert, we arrive at alternative models 1) a single amalgamated plane or a plurality of planes. Both of course a metaphorical. Neither requires mathematical transformation (no volume integrals to surface integrals are required to explain cosmic counter action). Euclidean space exists in the form of a volume when viewed with respect to an acceleration in one direction and as an infinite plane (or a plurality of parallel infinite planes when viewed in a direction perpendicular to the acceleration).

On the large scale, the universe is smooth, on the small scale it is lumpy. To formulate the cosmos as a unified whole, space and mass must be fully homogenized. Lumps of matter dead ahead (until contacted) do not exert a retarding influence. Matter in the opposite direction exerts a gravitational retarding influence upon the acceleration, but this diminishes inversely with the square of the distance. To apply the inertial resistance of the cosmos to an accelerating body '**B**' anywhere in the universe, the force issued by the cosmos is independent of the location of '**B**.' A universe comprised of thin planes will act orthogonally to collectively oppose acceleration equally at any location. Space and lumpy mass amalgamate uniformly within the concept of the infinite plane.

As shown in **Fig16**, the effect of distributed matter (M_o) along a line of action collinear with the accelerating motion of a body can only exert its presence by actual contact or through the influence of a field such as gravity which diminishes inverse square with distance. Because average Hubble density is vacuously small, the retarding force encountered while moving at a uniform rate, is insignificant. Acceleration, by contrast, is instantly opposed by the omnipresence of inertial space.

The constitution of *"inertial space"* as the acceleration impedance of the universe, takes into account only the operative form of inertial resistance. That area density can be formulated either as a single slab σ_U perpendicular to the direction of acceleration or as a plurality "n" of parallel planes having density σ_U/n, requires that 2-sphere scale be adjusted to take into account 3-sphere gravitational energy in the laminar construct. The same mass will be involved when the universe is considered as one area density or as a number of parallel area density planes spaced over the extent of the cosmos. No additional energy is considered created between the laminae themselves. To model the universe as area density, it must include 3-sphere gravity energy, but nothing more

Hubble mass energy is comprised of bare mass energy $M_b c^2$ plus negative gravitational binding energy U_3 of a 3-sphere having uniformly distributed bear mass M_B, that is[11]

$$\mathbf{M_3} = \mathbf{M_b} + \mathbf{U_3} = 2\mathbf{U_3} = \frac{6(\mathbf{M_b})^2 \mathbf{G}}{5\mathbf{R}} \tag{32}$$

Total energy of a 2-sphere is:

$$\mathbf{U_2} = \frac{(\mathbf{M_b})^2 \mathbf{G}}{2(\mathbf{R_2})^2} \tag{33}$$

---

[11] Bare mass is the sum of the masses measured by tearing the sphere into many small pieces and separating them to avoid any gravitational interaction

Transformation of Hubble Energy to Infinite Plane

Both Newton and Einstein opined that inertia could not be an antonymous property of the body undergoing acceleration, independent of some connection to the universe. But to admit a cosmological source within the predicate of Mach's Principle, was to Einstein, an effrontery to the Theory of Special Relativity. Herein, the mystery of inertial space as instantaneous communicator of distant matter is addressed by a double transformation of the Hubble from 3-sphere —> 2-sphere —> infinite plane as depicted in **Fig 16**. If the fully homogenized 3-D Hubble sphere of radius R_3 is transformed to a 2-sphere of the same radius, it would have less gravitational energy by a factor of **5/6**. To recover lost gravitational energy when modeling the universe in its functional form as an infinite inertial plane, one needs to adust the effective mass of the 2-sphere upward or lower the radius R_2. Herein, the ensuing gravitational energy deficit is redressed by reducing the 2-sphere radius R_2 by **5/6**. In the 2nd transformation, the 2-sphere is transformed to a flat plane, resulting in an additional gravitational energy deficit of 50%. For present purposes, R_3 will be provisionally taken as the Hubble scale $R_H = 1.3 \times 10^{26}$ meters, corresponding to **H = 70**. The corresponding Hubble mass $M_u \approx 1.5 \times 10^{53}$ kgm. The effective radius R_2 for purposes of calculating the area density of the infinite plane is approx **1.1×10^{26} meters**.

| Uniform Density Hubble 3-Sphere with G energy $U_3 = [3(M_u)^2][G/5R_3]$ | Transformation of 3-D sphere to 2-sphere with no energy change reduces R_2 | 2-sphere transformation to infinite flat plane reduces gravitational energy 50% |

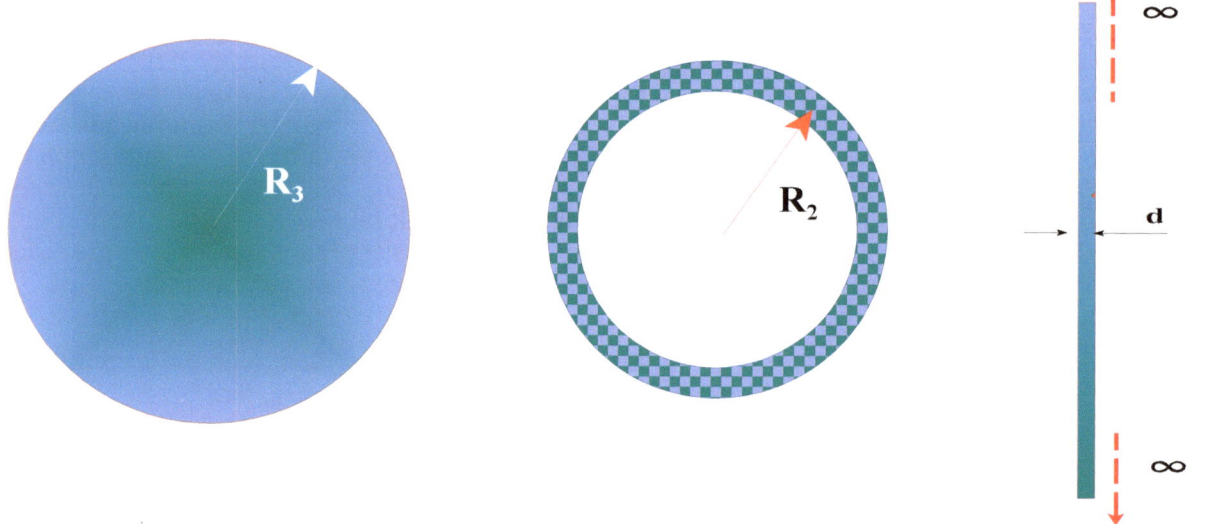

FIG 16

Physically, the Hubble is 3-D, but operatively it is a surface density σ_U rather than a volume density ρ_U. Intuitively, one might reason the transformation from 2-sphere to infinite plane could be carried out by simply setting $R_2 \rightarrow \infty$ (Assuming new volume adds mass proportionately, so density remains constant). But alas, that is not the way the world is made -- the force produced by an infinite plane is half that produced by the infinitely large 2-sphere geometry.[12]

---

[12] The infinite 2-sphere and infinite plane are mathematically different geometries. Half of 2-sphere energy is in the form of gravitational binding, which is lost when transforming from 2-sphere to infinite plane.

From and inertial perspective, the universe acts as an area density per Newton's 2nd Law. The problem posed is that of determining the inertial density operative of the Hubble universe in terms of Hubble parameters. Gravity energy $M_{g3} = [(3M_b^2 G)/5R_3]$ based upon a Hubble 3-sphere construct is larger than the area density $M_{g2} = [M_b^2 G/2R_2]$ built from scratch using the bare mass M_b spread over a 2-sphere shell of radius R_3. The effective radius R_2 can thus be determined from (32 and 33):

From which

$$U_3 - U_2 = \frac{3M_b^2 G}{5R_3} - \frac{M_b^3 G}{2R_2} \qquad (34)$$

$$R_2 = (5/6)R_3 \qquad (35)$$

The unavoidable interpretation of (35) is that the Hubble sphere cannot exists as a 2-sphere shell area density while at the same time maintaining its 3-sphere status of radius R_3. As a 3-sphere, the gravitational energy U_3 is greater than the gravitational energy U_2 of the 2 sphere construct. If the universe is to act as an area density (as it must to explain inertial reaction in terms of an infinite plane), the only energy available is that which corresponds to its operational mode (at this stage, the 2-sphere interim construct will have considerable negative gravitational energy U_2). For purposes of determining 2-sphere area density and flat plane energy density, G and bare mass are provisionally considered fixed factors. That the 2-sphere has the same total energy as the 3-sphere while acting as a 2-sphere area density, the operative value of R_2 must be reduced to $(5/6)R_3$ per (35). The total energy of the Hubble universe as a 2-sphere operative area is the therefore:

$$E_T = \frac{M_b c^2}{1} + \frac{M_b^2 G}{2R_2} = \frac{M_b c^2}{1} + \frac{M_b c^2}{4\pi R_2^2 \sigma_U} \qquad (36)$$

At this juncture, the 2-sphere view of the Hubble sphere as an interim configuration state, would be characterized as having a total energy:

$$E_T = Mc^2 + U_2 = Mc^2 + \frac{M_b^2 G}{2R} \qquad (37)$$

Where in a zero energy universe, positive matter energy Mc^2 equals negative gravitation energy U_2, in which case, one might assume an infinite radius spherical shell would produce the same result as an infinite area flat plane having the same area density. But a shell universe has zero internal gravitational force whereas the infinite flat plane is deemed to provoke an equal gravitational force of infinite extent in opposite directions normal to the plane. For present purposes, the gravitational implications can be disregarded, whence the operative mass and effective area are determined by way of Gauss's divergence theorem. That there will always be two diametrically opposite operative areas to be considered for any volume completely enclosed by a Gaussian surface, we arrive at the effective density E_{ED} of the flat plane befitting a Hubble sphere of scale $R_H = R_3$ reduced to an effective radius $R_2 = (5/6)(R_3)$ and mass energy equal to the bare mass M_b. As an inertial plane σ_U is distinguished by the total absence of gravity energy created by the action of gravity acting upon gravity. Based upon galactic counts, an estimate of bare Hubble mass $M_b = 1.5 \times 10^{53}$ kg, whence:

$$\sigma_U = \frac{M_b}{4\pi R^2} = \frac{1.5 \times 10^{53} \text{ kg}}{12.56 \times (1.1 \times 10^{26} \text{ meters})^2} \approx \frac{1 \text{ kg}}{\text{meter}^2} \qquad (38)$$

The gravity energy of the shell universe (per 37) goes to zero when $R \rightarrow \infty$. For $R = R_2$, the gravitational spatial energy for a mass M_B defined by a Hubble sphere of radius $R = R_2$ is obtained by imagining the reactionary acceleration uniformly spread of the Hubble shell as shown in **Fig 17**.

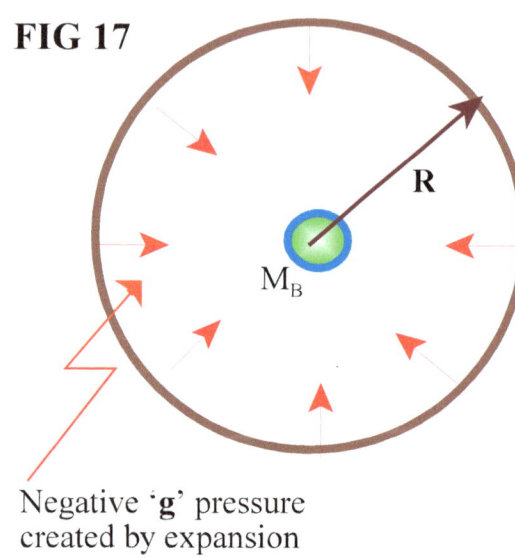

FIG 17

Negative '**g**' pressure created by expansion

$$E_S = -3PV = -3\left[\frac{a_n M_B}{4\pi R^2}\right]\left[\frac{4\pi R^3}{3}\right] \quad (39)$$

and for $a_n = c^2/R$, then

$$E_S = (-)\left(\frac{c^2}{R}\right)M_B R = M_B c^2 \quad (40)$$

Negative energy in the spatial '**g**' field of M_B is equal to the positive $M_B c^2$ mass energy.

What is shown, is that for an acceleration factor $a_n = c^2/R$, the '**g**' field energy of a mass is always equal and opposite to its positive Mc^2 energy. That this is the condition for a zero energy universe, attention is called to the gravitational equations first synthesized from the Theory of General Relativity (circa 1922).[13] Here we are concerned only with what is known as the 2nd Friedmann equation which includes the cosmological constant Λ:

$$\ddot{R} = -\frac{4\pi G}{3}\left[\rho_u + \frac{3P_s}{c^2}\right]R + \frac{\Lambda R}{3} \quad (41)$$

When $\rho_U/c^2 = -3P$, net energy is zero and (41) simplifies to:

$$\ddot{R} = \frac{\Lambda R}{3} \quad (42)$$

Using Einstein's value for the cosmological constant ($\Lambda = 3H^2$), we see that de Sitter's empty universe and the zero energy universe (Positive Mc^2 energy = negative gravitational energy) have the same solution:

$$R = R_o[e^{Ht}] \quad (43)$$

Ergo, a zero energy universe expands exponentially. And per (40), negative '**g**' pressure created by exponential expansion ($a_n = c^2/R$) will be equal to the positive Mc^2 energy that created the '**g**' field.

---

[13] Known as Friedmann-Lemaitre equations after the two men who independently discovered them (Alexander Friedmann and George Lemaitre).

From (40), it is at last possible to formulate the constancy of the **MG** product, the imperative upon which orbital stability rests. The common opinion that **M** and **G** are independent invariant factors can most likely be traced to the supposition that **M** involves an assembly of real atoms (i.e., matter), which do not change in number and G has for over three centuries retained its title as Newton's gravitational constant. In reality it is an acceleration factor having mks units 6.7×10^{-11} m^3/sec^2 kgm^{-1} Why **G** should have this particular value of volumetric acceleration per unit mass and why it should be a constant in an evolving universe, has always been a mystery. In spite of great efforts to ensure its durability as a constant by some, it has long been suspected by others to be a variable that decreases inversely as the Hubble scale **R** increases.[14] The prospect of variable **G** creates havoc for theories founded upon its invariance.

While the many attempts to measure drift proved fruitless, it has been stressed in the author's other works and elsewhere, that all such experiments were based upon the proposition that inertial matter is also constant. The nil result was imputed by the keepers of the standard model, as proof of constancy. There is no law of conservation of mass. Mass is a form of energy. Energy is the conserved quantity. Matter seems to be permanent, even though it is known to be otherwise.

In this sense it is easy to unintentionally find oneself in the apocryphal state of resisting the idea that the inertial property of a given amount of atoms can vary. Masses have gravitational fields; the volume of the gravity field of a fixed quantity of atoms will increase as the universe expands. An increase in the volume of the negative energy 'g' field of a mass will increase the inertial property of the mass. The inertia of a mass **M** is determined by the energy contained in the 'g' field of **M**, not by the quantity of atoms that make up the physical construct. To oppose acceleration, the volume of the 'g' field of **M** interacts with the inertial impedance σ_U of the universe. Thus, while, varying **M** and varying **G** are too much of a shake-up for conventional proclivity, it is the "*Standard Model*" that fails to explain inertia and gravity, nor does it offer an answer to the question as to how an expanding, evolving universe invents, implements and retains the value of **G** through all eternity? Herein, Inertia '**M**' and Gravity '**G**' are determined to be reciprocal interdependent(s). From (39):

$$E_S = -3PV = -3\left[\frac{a_n M_B}{4\pi R^2}\right]\left[\frac{4\pi R^3}{3}\right] = \frac{a_n M_B R}{1} \tag{44}$$

and since from (17)

$$G = \frac{a_n}{4\pi \sigma_U} \tag{45}$$

Then

$$GM_B = \frac{a_n M_B R}{4\pi \sigma_U} = K \tag{46}$$

---

[14] Dirac's large-number hypothesis (**LNH**). Reasoning that the near equality of the *electro/gravitational* force ratio to the *cosmic/subatomic* size ratio was more than a coincidence, Dirac opined that these large numbers will maintain the same proportions at all times. As a consequence, the value of **G** would diminish inversely with the size of the Hubble universe. The general physics community nonetheless, discounted the LNH and other variable **G** theories based upon the observed stability of orbits in the solar system. Apparently unconsidered was the fact that orbital parameters depend upon the **MG** product, rather than **G** alone.

The Flat Earth Model

From (46), **MG** field energy is independent of **R**. Because **G** diminishes inversely with **R**, inertial mass **M** must augment as **R** increases. Ergo, the **MG** product is constant. What is revealed by (44), (45), and (46), is that the orbital path of a satellite about a much larger central mass **M**, will be stable, not because '**G**' is constant, but because the **MG** acceleration field is constant.

From the perspective of a unidirectionally accelerated body '**B**', the universe is seen as a metaphorical flat plane density $\sigma_U = 1$ **kg/m²**. Inertial field lines are parallel to each other and perpendicular to σ_U. The same body '**B**' perpetually experiences the isotropic divergence of expanding space. If **B** is a uniform density spherical object, the inertial reactionary field lines will be radially convergent upon the surface of **B**. In both cases, the observed force is a secondary effect, the inevitable import of the Einstein's doctrine of relative acceleration. All of the known physical forces have dynamic foundations.

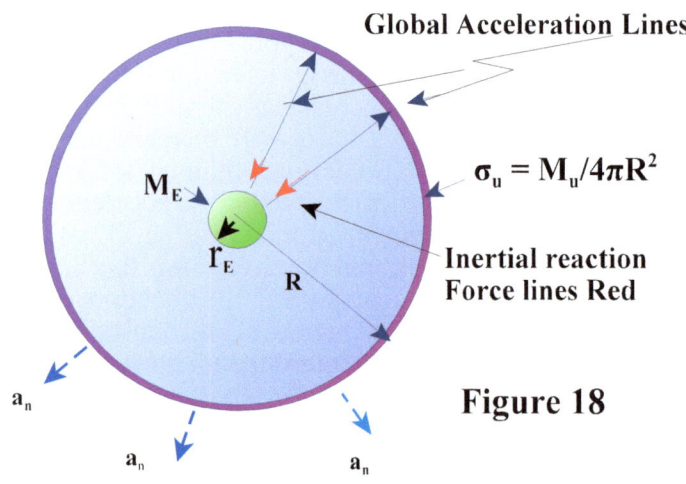

Figure 18

Fig 18 depicts the earth as the center of its Hubble sphere. The divergent spatial acceleration field (black arrows) acting upon mass M_E creates an isotropic inertial reactionary force in accordance with Newton's 2nd law. To calculate the reactive force at the earths surface, the Guassian surround is fitted to have a radius r_E equal to the earth. Accordingly, from **Figure 4**, and Newton's second law,

$$\frac{F}{A} = \left[\frac{M}{A}\right]a = P = \sigma(a) \tag{47}$$

Then for the earths reaction to the cosmological acceleration a_n, the pressure is:

$$P_E = (M_E/A_E)(a_n) \tag{48}$$

And for the reaction of the universe due to earth's counter reaction,

$$P_U = (M_u/4\pi R^2)(a_E) \tag{49}$$

The two pressures must be equal since there is no net flow across the space mass interface, and therefore:

$$(M_E/A_E)(a_n) = (M_u/4\pi R^2)(a_E) \tag{50}$$

Fig 19, The inertial reaction a_E of the earth's inertial mass at its surface is '**g**,' and therefore:

$$\frac{g}{a_n} = \frac{\sigma_E}{\sigma_u} = \frac{\dfrac{M_E}{4\pi(r_E)^2}}{\dfrac{M_u}{4\pi(R)^2}} = \frac{\dfrac{M_E}{4\pi(r_E)^2}}{\dfrac{kgm}{meters^2}} \qquad (51)$$

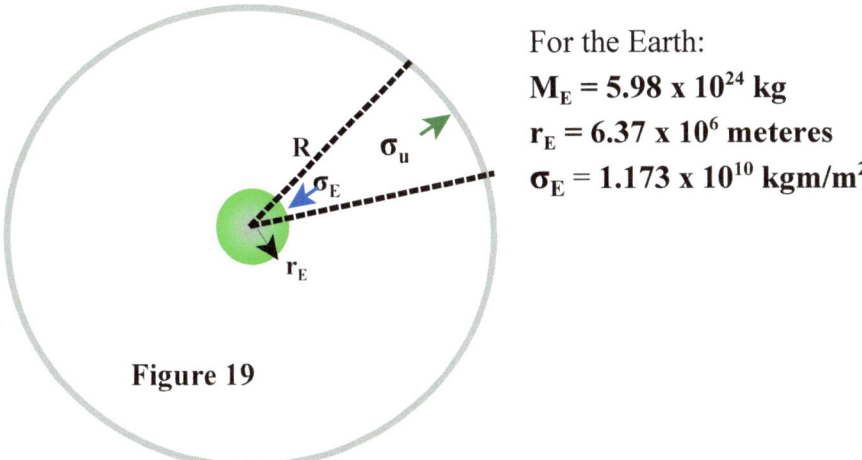

For the Earth:
$M_E = 5.98 \times 10^{24}$ kg
$r_E = 6.37 \times 10^6$ meteres
$\sigma_E = 1.173 \times 10^{10}$ kgm/m²

Figure 19

Where $a_n = c^2/R$ [a q = -1 exponentially expanding universe], the earths '**g**' field at its surface is:

$$g = [(\sigma_E)/(\sigma_u)]a_n = 9.8 \text{ m/sec}^2 \qquad (52)$$

The close correspondence between the empirically determined size, mass and gravity at the earth's surface adds credence to our **kgm/m²** estimate of scalar reactive density. That the "*standard international unit of inertial mass to metric area,*" must be exactly "**1**" we revisit (14) and the development thereof. Hence we can than again confidently express **G**

$$G = \frac{(-q)3H^2}{4\pi\rho_u} = \frac{-q(3H^2)}{4\pi\left(\dfrac{3\sigma_u}{R}\right)} = \frac{-qc^2}{4\pi R\sigma_u} \qquad (53)$$

Using the same cosmic sigma value σ_u calculated earth's reactionary '**g**' field (equation 22), there results:

$$G = 6.67 \times 10^{-11} \text{ m}^3/\text{sec}^2 \text{ per kgm}. \qquad (54)$$

Both Big **G** and local **g** fields depend from the one **kgm/m²** scalar density cosmic modulus. Likewise, so also does inertial reaction. This should not be surprising in that each depends upon the other. Specifically, the earth's reactionary "**g**" field (22) derives from **G**, and **G** depends from isotropic cosmological acceleration c^2/R. A Gaussian surround fitted to the earth-space interface reveals the presence of a negative pressure acceleration field ["force-per-unit-area"] acting upon M_E.

Figure 20: Gravity is founded upon "*Second Law Symmetry.*" The emergence of reactionary force fields follow straight away as manifest of density difference at the space mass interface r_E. Isotropic spatial expansion carries with it the inertia imperative σ_U - and therefore 'g' field(s) proportional to the inertia **M** of the masses. M_E represents the earth's mass, r_E its radius. The affect of M_e is the effect g_E, the earths gravitational counter reaction field $M_E(a_n)$ at the radius r_E. Taking the earth as a spherically uniform density of radius r_E, the counter pressure at all points of the surface $4\pi r_E^2$ must equal the source pressure that produced the counter action. Specifically,

$$\frac{F}{A} = \frac{M_u}{A_u}(g) = \frac{M_e}{A_e}(a_n) = \sigma_u g \qquad (55)$$

Equation (55) is Newton's 2nd law in gravitational field form: "*The pressure created by the cosmological acceleration field a_n acting upon a local mass having a mass to area ratio M_e/A_e is equal to the pressure created by the 'g' field counter acceleration acting upon the cosmological scalar density function σ_u.*"

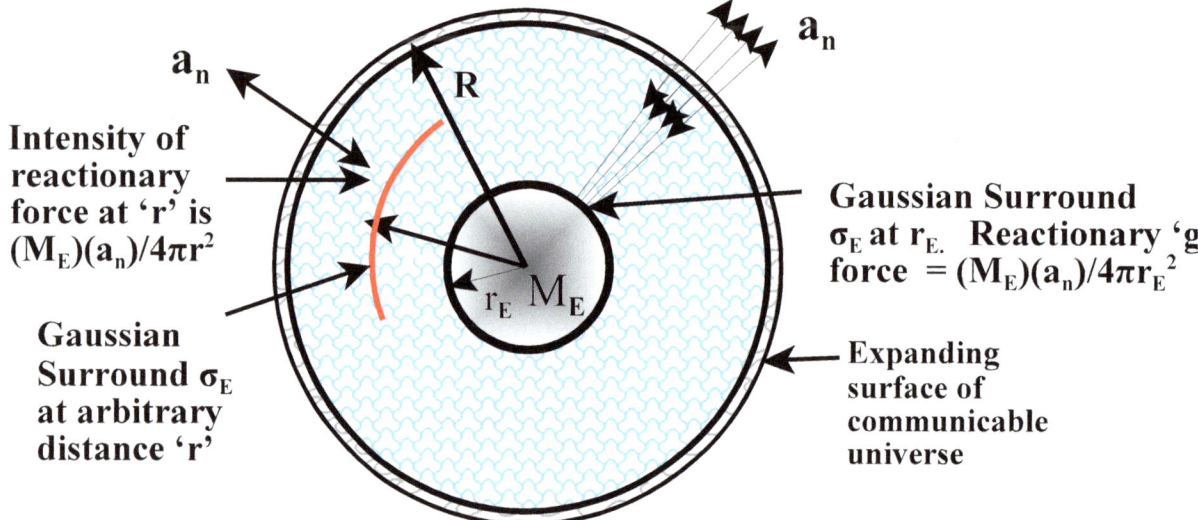

The force intensity per unit area at a distance **r** beyond r_e is inversely proportional to the surface area $4\pi r^2$ of an imaginary concentric sphere of radius **r** over which the force is distributed. Thus, for any distance 'r' from the mass center ($r_e < r < R$), reactionary counter acceleration will be proportional to the isotropic acceleration field divided by the area of the imaginary Gaussian sphere $4\pi r^2$.

The workings of the universe cannot be explained by the present state of conventional proclivity. To understand the cosmos, one must take notice of inertial space and the significance of size. One cannot tinker with one item alone - to get it right, requires that heretofore constants simultaneous vary. A change in one means change in another – notice hearin below the following:

1) Exponential expansion is predicted by the zero energy universe, equation (41). There is only one solution to the Friedmann equations for a zero energy universe based upon GR. No dark energy is required nor is any allowed.

2) The Expansion Field of space is one-in-the-same as that which causes gravity.

3) Einstein's cosmological constant Λ, originally introduced as having a value $3H^2$ such that, when multiplied by $R/3$, it would prevent gravitational collapse - in the end, is revealed as the $q = -1$ exponential expansion rate c^2/R, which turns out to be the cause of gravity.

4) The universe appears critically balanced at omega = 1. The misinterpretation of gravity as an independent force leads to the fallacious doctrine of finely tuned constants. That G encodes the cosmological acceleration rate c^2/R, no fine tuning is required. The universe will always appear to be delicately balanced between expansion and gravity because gravity is the result of expansion.

5) The Hubble mass is $4\pi R^2 \sigma_U$, increasing as it does with the square of the scale factor R, the inertial density is always = 1 kg/meter^2. No new, particles or dark energy is involved.

6) The MG product of an individual body is always constant. G decreases as $1/R$ and M increases proportionally with R.

7) The universe is observed as a dilating sphere, but it acts functionally as an infinite plane having unity density = **1 kg/m^2**.

8) The effective Hubble scale for computing area density is **R $\approx$ 1.1 x 10^{26} meters**

9) The bare mass of the Hubble is $\approx$ **1.5 x 10^{53} kg**.

10) The herein developed complexion of the inertia/gravity complex completes the analytical construct of the two eminent theories of gravity. Newtonian and Einsteinian formulations can now be understood as emergent inertial reactions predicated by global expansion.

"There is nothing more difficult in hand in the introduction of a new order of things, because the innovator has for enemies all those who have done well under the old condition, and lukewarm defenders in those who may do well in the new."

NICCOLO MACHIAVELLI
Il Principe 1513

///

///

www.ingramcontent.com/pod-product-compliance
Lightning Source LLC
Chambersburg PA
CBHW041257180526
45172CB00003B/881